The Development of India's Forest Resources

The Development of India's Forest Resources

Compiled by

ECONOMIC BRANCH OF THE FRI, DEHRADUN

BIOTECH

2018

BIOTECH BOOKS®

ISBN 978-81-7622-073-6

Published by: **BIOTECH BOOKS®**
4762-63/23, Ansari Road, Darya Ganj,
New Delhi - 110 002
Phone: +91-011-23262132
E-mail: biotechbooks@yahoo.co.in

PRINTED IN INDIA

Photo by I. N. Sharma.

Log Pond and Gantry
(The floats indicate sunk logs).
New Forest Research Institute, Dehra Dun, 1924.

INTRODUCTION.

Compared with other parts of the British Empire, India is in an enviable position in having introduced a policy of forest conservation prior to the destruction of her extensive natural forests. The policy of saving from destruction or alienation as large an area as possible was inaugurated in the middle of the last century and came at an opportune time, in advance of the rapid development which followed on a settled rule. In some other countries settlement on the land has involved the destruction of the natural forests and it has only been after the realisation that destruction had proceeded too far that a policy of conservation was introduced.

2. It is unnecessary to justify the policy of Forest Conservation. Forests represent a form of wealth which can only be fully maintained by the State. Few individuals are in a position to invest in a form of capital which can only be realised after a long interval of time. The peculiar nature of the forests necessitates their constitution and protection under special laws. It is essential also to make a close study of the conditions of growth to ensure that the soil will attain its maximum productivity. This falls within the province of silviculture and forest management but it is equally necessary to ensure, after having attained the desired degree of productivity, that the products of the forests be utilised to the best advantage. The objects of management must of necessity vary with local circumstances, in some cases the forests are required to supply the domestic requirements of the local population, including pasturage; or they may be managed for purely commercial purposes. It falls within the province of the department charged with the administration of this large government estate to strive continually to attain the maximum output in materials or revenue or both, the policy being indicated by the objects of management.

3. It is usual to measure the success of forest management entirely according to financial results. In a country like India, in its present state of development, it is not justifiable to make this the sole criterion. The uses of forest produce for certain purposes are already well established but it is not the case that all the available produce is being utilised or being made the best use of. There is scope for the creation of many new industries, all tending to replace imports, and at the same time to find employment for indigenous enterprise, capital and labour. It is the aim and object of forest research to develop the resources of the country. The increase of revenue is incidental and by no means the most important benefit which accrues. The inci-

dence of royalty on forest produce is usually a relatively small proportion of its market value and in the case of products which have to be collected and transported at a heavy cost, and perhaps manufactured by expensive machinery under costly supervision, the initial royalty or value of the raw material in the forest is relatively small. The benefit to the country arises in the money which changes hands in the process of handling, manufacture and marketing. For these reasons therefore the criterion of the revenue actually obtained is not the real test of the forest estate's true functions or of the efficiency of its management.

4. That there is endless scope for extension in the utilisation of forest products in India is undeniable. There is equal opportunity for sound enterprise but enterprise to be sound must be based on and guided by solid facts. The department has endeavoured, by the agency of a well equipped Research Institute, to collect and make known those facts which are all-important to the next stage of economic development, namely, the commercial or semi-commercial attempt to develop an industry based on forest products.

5. Three stages of research are usually necessary. The first is the purely scientific enquiry, usually ending at the laboratory stage. The second involves an inventory in the forest, to ensure maintenance of supply of the raw material together with the manufacture, on a scale sufficient to test the market value and suitability of the materials or products in question. If this fails to establish a demand the third stage, comprising the erection of pioneer plants or factories, perhaps on a limited scale, will be necessary.

6. The research which is now being done aims at covering the first and in certain directions, *e.g.*, paper pulp, the second stage. In the following pages the research officers connected with the utilisation of forest products have given an account of the nature and present scope of their work.

The field is large and, as yet, only the most important sections of the work in hand have been inaugurated. A great impetus was given by the war, at a time when India was supplied with difficulty from outside. The hiatus in industry which has followed has afforded an opportunity, in the sections herein described, of making up some leeway; but research, to be productive of the best results, must not only march with, it must anticipate the public demand. Our existing efforts therefore leave room for much expansion but the very nature of forests and the character of their products compel slow and patient progress. Few of us live to see the results of our work in the forests and in the testing of timbers for specific purposes time again is the main factor. Results must not therefore be expected too soon. At the present moment, while much positive information and advice can be given, at least an equal proportion of negative information is available. If the latter can help in preventing serious risks or errors in commercial enterprise its utility is not wasted.

7. The value of the work dealt with in this Note depends on the degree of confidence engendered in the public and in the Department outside the Institute. The general impression is that the results so far achieved are gratifying and fully up to expectations. The staff has therefore every reason to be proud of the beginning which it has made.

W. F. PERRÉE,

President, F. R. I. & College, Dehra Dun.

PREFACE.

This Note on the work of the Economic Branch of the Forest Research Institute is presented with a two-fold object, namely, to convey to persons interested in Timber and Minor Forest Products, some idea of what is being done to place forest produce on the market in a suitable form and secondly to put enquirers in a position to judge as to what extent their enquiries can be answered and what value to place on information supplied.

"India's Forest Wealth," a recent publication by Mr. E. A. Smythies, brings very clearly to the notice of the public, what products are available from the Forests of India, so that this summary of the methods employed to develop the resources of our forests may be considered to be a corollary to the above mentioned publication.

The subject matter is presented in eight sections, the first of which deals with the history of the Economic Branch, which it is hoped may be of general interest. This is followed by a short note on the activities of the Section of Wood Technology. Section III deals with Timber Testing, and has been prepared by Mr. L. N. Seaman, the Officer-in-Charge, who may be said to have inaugurated Timber Testing in India on really scientific lines, having formerly been in charge of the Canadian Dominion Government's Testing Laboratory at Montreal. The Section on Seasoning, No. IV of this Note, is by Mr. Stanley Fitzgerald, formerly Adviser to the Air Ministry and Admiralty on Kiln Seasoning, and now in charge of this Section at the Forest Research Institute. The Section on Wood Preservation is by Mr. J. H. Warr, who has had long experience in the antiseptic treatment of timber, both on a commercial and laboratory scale. Section VI, on Paper Pulp, is by Mr. W. Raitt, a paper pulp expert of long experience, who is sufficiently well known both to the public and to the paper world to require no further introduction. The Section on Minor Forest Products and Sub-Sections on Wood Working, Veneer and Sawmills are by Mr. H. Trotter, Assistant Forest Economist, a Burma Forest Officer, who has had several years experience of the Economic Branch.

RALPH S. PEARSON,

Forest Economist.

The Development of India's Forest Resources.

I.—Brief History of the Economic Branch.

BY

R. S. PEARSON, *Forest Economist.*

"It may safely be said that there is hardly any Government Department in India whose work and aims are so little realized by the general public as are those of the Forest Department."

The above is what Prof. R. S. Troup, C.I.E., wrote in 1912 in his introduction to the "Work of the Forest Department in India" which is a brief resumé of the activities of the Department and well worth perusal.

If little is known of the work of the Forest Department by the general public, acquaintance with the Economic Branch of the Forest Research Institute must be very limited, in fact probably most people have never even heard of its existence.

Small boys, small men, small institutions and even small nations like occasionally to come under the lime light and as the work carried out in the Economic Branch is thought to be really interesting and is already of proved value in developing the forest resources of British India and assisting in the creation of new industries, a short account of what is being carried out and results obtained to date may be of interest. Moreover, such research work costs money and it is only right that the taxpayer should know how this money is being spent.

In 1906 a definite scheme was prepared and sanctioned to establish a Forest Research Institute at Dehra Dun. The original idea was that of the late Mr. R. C. Wroughton, Inspector General of Forests, who before retiring broached the subject to the Government of India. It was however, to Sir Sainthill Eardley-Wilmot, his successor, that the credit is due for starting and organising the Forest Research Institute. Five branches were created, namely those of Silviculture, Forest Entomology, Forest Chemistry, Forest Botany, and Forest Utilization, commonly known as the Economic Branch.

The first officer appointed as Forest Economist in 1906 was Mr. R. S. Troup, now Dr. Troup, Professor of Forestry at Oxford University. In 1909 he took over charge of the Silvicultural Branch and the writer succeeded him as Economist and, with the exception of short periods of absence on leave, has held charge for the last fifteen years.

Forestry as a subject has many branches. Broadly speaking it involves silviculture, the management of woods, the preparation of working plans, protection, exploitation and finally, utilization of the produce. At the time when scientific forestry was introduced into India control of any sort was virtually absent, and the limited staff available was engaged in exploring, demarcating, and carrying out the settlement of the existing forests, and thus the work gradually extended to protection and to more careful management, which primarily aimed at restoring the much damaged forests. Later, the work intensified and working plans were gradually introduced, but what of utilizing the timber and minor forest products, such as tan stuffs, cutch, lac, gums and resins and a host of similar articles ? Little or nothing could be done to develop the vast resources of the State forests, further than to sell such species of timber as were already on the market, as the staff was not sufficient and their other work was too heavy to permit of serious attention being paid to anything but the best and easiest saleable woods. It was not until Mr. J. S. Gamble, M.A., C.I.E., F.R.S., F.D.S., a most distinguished Conservator of Forests, commenced in 1877 to carry out a long and careful examination of a very large number of our Indian timbers, that real interest began to be taken in the possibilities of more than about two dozen of the best known woods. That officer published a book, which has now gone through several editions, entitled "A Manual of Indian Timbers ", which still remains our standard work.

It may safely be said that the more intense utilization of our woods is primarily due to Mr. Gamble's careful and prolonged studies, of which his Manual stands as a record. The position in 1906, when the post of Forest Economist was first sanctioned, was therefore, that the most important literature available on Indian Timbers was to be found in Gamble's Manual, supplemented by articles which had from time to time appeared in the " Indian Forester." Existing Floras also referred occasionally to the uses of timber.

Important literature available on minor forest products was to be found in Gamble's Manual, Watt's Dictionary of Economic Products, Talbot's List of Trees and Shrubs, etc., of the Bombay Presidency and Kanjilal's U. P. Flora ; while in 1909 the post of Reporter on Economic Products was abolished and the records which dealt with Forest Products and which had been collected by Sir George Watt and his successors were transferred to the Forest Economist's office.

The Research Institute was originally evolved from the Forest College and an attempt was made to merge the staffs of the two together. At the outset work was commenced in such space as was available in

Photo by T. B. Chitrakar.

The original workshop of the Forest Economist, 1907-1913.
Forest Research Institute, Dehra Dun.

[To face page 3].

existing buildings and the school museums formed the nucleus of the research collections.

A fair collection of minor forest products was available, arranged with a mass of botanical, geological and other exhibits, while Gamble's most valuable hand collection of timbers, on which he based his Manual, was fortunately still in fair condition, though not suitably arranged for reference purposes. Of laboratory equipment there was none, nor were any buildings available, in which to erect experimental plant, in fact the only available space was a small office for the Forest Economist and a somewhat larger room for his clerical staff and records. It must also be remembered that there was no correspondence, as no connecting link had been established between the Economic Branch and other forest officers in the provinces or with the commercial world.

The establishment consisted of one Imperial Officer, two clerks and a collector, surrounded by a considerable number of miscellaneous files and collections all of which had to be sorted out and classified. The first work undertaken by the Forest Economist was to classify the available literature and to start permanent Record Files. Later, in 1909, the collections were taken in hand, added to and a fair Minor Forest Products museum arranged in one of the halls of the Forest College. There was no space available for establishing a good timber exhibit, and the best that could be done was again to arrange Gamble's hand collection of timbers and house it in a verandah. In the meantime the files of the Reporter on Economic Products came to hand and required sorting out. This entailed a heavy task and an Assistant Forest Economist was appointed to undertake the work, on which he was employed for over 18 months.

The first enquiry undertaken was that on possible match woods, and owing to want of equipment the woods to be tested had to be sent to Europe. On this enquiry Prof. Troup issued the first Forest Memoir pertaining to Forest Utilization. In 1909 a very extensive enquiry on wood preservation was inaugurated, chiefly in connection with sleeper woods and this at once brought up the question of a laboratory or workshop. As has been stated above none was then available, as there was no money, or to be more correct, the confidence that money spent would yield any return had not at that time been established. The difficulty was, however, partially overcome by the Economic Branch appropriating an old shed or godown belonging to the College, in which firewood was stored and which came in useful for heating the antiseptics with which the small specimens were treated. The photograph of this shed on the opposite page may be compared with the present more extensive laboratories and workshops recently erected.

As the sphere of activities widened, new investigations were started. the most important of which dealt with a very extensive enquiry into the utilization of bamboos for the manufacture of paper pulp, the development of the sleeper enquiry, the question of more up-to-date methods of distilling essential oils, the development of the utilisation of *Boswellia serrata* gum-oleo-resin, and many others which are dealt with under their respective headings later on in this note. It is sufficient to say that the work increased and, what was even of greater importance, a definite connection was established with the local forest officers, the commercial world and the railway engineers.

The justification for further expenditure and the expansion of the Forest Research Institute became more and more apparent, and the next definite step came in 1914, when it was decided to erect a large central building, surrounded by what was then thought to be suitable laboratories and workshops. Of the accommodation available in the main building two large museums were allotted on the ground floor to the Economic Branch, one for the timber specimens and the other for Minor Forest Products. The wood museum was sufficiently large to permit of Gamble's hand collection being suitably arranged and a fine collection of large size specimen planks and other interesting exhibits being displayed to advantage. The adjoining photograph is of the wood museum; looking down one side it shows the larger specimen planks to the right, stands containing Gamble's collection to the left, while gun-carriage wheels made of Indian timbers and many other exhibits may be seen in the centre. The Minor Forest Product collections which have been more than trebled, have been suitably catalogued, and arranged in new show cases in another museum.

The workshops and laboratories erected round the main building consisted, as far as the Economist Branch was concerned, of a timber testing workshop, where other experiments were also carried out, an experimental timber treating plant, working according to the open tank method, and storage room or godowns. Later a small wood workshop, and a seasoning kiln of a temporary nature, were added.

At the time this new Institute building was erected it was anticipated that the accommodation would be sufficient without further expansion for 20 years at least. War, however, taught many lessons and amongst them it taught research officers, that their staff, equipment and accommodation was entirely inadequate to meet the numerous enquiries and investigations which had to be undertaken. The war, moreover, gave a much needed impetus to the utilization of the various timbers found in the forests of India and Burma, other than those in common use, and this fact, more than any other, drove home the importance of further expansion in that branch of the Institute which was primarily concerned with wood utilization.

In 1918 schemes were considered to develop the present Institute and these ideas were generally accepted and approved early in 1919

Photo by I. N. Sharma.

Economic Museum.
Forest Research Institute, Dehra Dun, U.P.

[To face page 4].

at the Triennial Conference of Conservators from all provinces of British India. The first idea was to expand the present Institute by taking up more of the adjoining land, but on going into details and keeping in mind the lessons learnt from the 1914 expansion scheme, it was finally decided to start a new Institute some two miles outside Dehra Dun and to utilize the present buildings for educational purposes.

The new scheme received the sanction of Government and the Secretary of State in 1920 and so as to meet all possible expansion for many years, both of research and education and including a demonstration forest, no less than 1,300 acres of land were acquired.

The difficulties in such a scheme, as far as the Economic Branch was concerned, were not the buildings, which could be designed and built to meet the requirements of the case, but the purchasing of suitable machinery of the right type and the still more difficult problem was now to obtain the services of first class specialists in each branch of the work. Most people would think that with a large and trained forest service to select from, the staff would be the easiest side of the business to settle satisfactorily, but then it must be remembered that the work of research in forest utilization is primarily a specialist's business, as forest officers only have to learn the rudiments of timber testing, wood preservation, kiln seasoning of timber, wood technology, veneer and wood working, and on the minor forest product side, paper pulp, tans, lac, etc., which are all specialised subjects and the studies of a life time.

To overcome these difficulties the Forest Economist was deputed by Government to proceed to England, Canada and America to study the lines on which similar work was being carried out in those countries, to purchase plant and to recruit staff. The Officer-in-Charge of the Paper Pulp section was also sent to England to purchase an experimental pulp and paper plant. By the end of 1920 some of the experimental plant and staff had arrived in India, but the new laboratories and workshops were not ready, and so it was decided to erect a small portion of the plant in the old laboratories, so as to get the specialists at work, and especially to train a nucleus of wood workers, and machine and kiln operators. Towards the end of 1921, the new laboratories and workshops were put in hand, and their construction made rapid progress. Erection of machinery began in 1922, but it took much longer than was anticipated and it was not until early in 1923 that the wood workshops and saw-mill were running and the timber testing laboratories in working order. The next section to start work was the experimental wood preservation pressure plant which opened in April 1923, while the sections of timber seasoning, paper pulp and veneers, which were held up by the lack of a water supply, have only been running since March 1924.

ECONOMIC

Forest Economist.—

Assistant Forest Economist.—

Section of Minor Forest Products.	Section of Wood Technology.	Section of Paper Pulp.	Section of Tans.	Section of Seasoning.	Section of Timber Testing.
Officer-in-Charge. MR. R. S. PEARSON (Officiating).	*Officer-in-Charge.* DR. H. P. BROWN (at Syracuse University U.S. A.).	*Officer-in-Charge.* MR. W. RAITT.		*Officer-in-Charge.* MR. S. FITZ-GERALD.	*Officer-in-Charge.* MAJOR L. N. SEAMAN.
		Imperial Assistant.		*Imperial Assistant.*	*Imperial Assistant.*
	Lab. Asst. . 1	Mr. M. P. Bhargava		Dr. S. N. Kapur.	Mr. C. R. Ranganathan, I.F.S.
					Upper Grade Assistant.
		Lab. Asst. . 1		Lower Gr. Assts. 2	Mr. Syed Mohammad Husain.
		Machine operators . . 2		Kiln operators . 6	
		Mechanic . 1		Kiln attendant . 1	
				Kiln cleaner . 1	Lower Grade Assts. 2
		Lab. attendant 1			
				Do. (D.L.) . 1	Head Computer. 1
		Peons . . 3			
				Peon. . . 1	Asst. Computers. 7
		Coolies (D.L.) . 6			
				Coolies (D.L.) . 3	Machine Operators . . 14
					Moisture determinaters . 6
					Machine cleaners 4
					Peons and Workshop coolies . 4
	TOTAL . 1	TOTAL . 14		TOTAL . 15	TOTAL . 38

BRANCH.

Mr. R. S. Pearson, C.I.E., I.F.S.

Mr. H. Trotter, I.F.S.

Section of Wood Preservation.	Section of Wood Working.			Sub-Section of Mechanical Engineering.	Clerical and Other Staff.
	Saw-mills.	Wood Work-shops.	Veneers.		
Officer-in-Charge.		*Wood Workshop Superintendent.*		*Mechanical Engineer.*	
CAPT. J. H. WARR.		MR. W. NAGLE.		MR. RAM DASS TANDAN.	
Imperial Assistant.					Head Clerk . . 1
MR. S. KAMESAM.					2nd Clerk . . 1
					Asst. Clerks. 17
				Electric mechanic . 1	Store Keepers . 2
		Foreman 1		Mechanics including Mistries and Fitters 12	Draftsmen . . 3
Plant operator 1		Carpenters (D.L.) 36			Painters . . 1
				Boilermen . . 5	
Peons . . 4		Coolies (D.L.) . 18			Motor drivers . 2
Coolies (D.L.) . 4		Other men (D.L.) 2		Mason and Black-smiths . . 3	Telephone operator 1
				Firemen, Motormen, etc. . . . 17	Peons, Chowkidars, Malis, Sweepers, Coolies and others 53
				Store Asst. . . 1	
				Khalasis . . 12	
Total . 9		Total . 57		Total . 51	Total . 81
					Grand Total .266

II.—The Organisation and Staff of the Economic Branch.

As is stated elsewhere the Economic Branch is one of five which together constitute the Forest Research Institute. The organisation in 1907 consisted of the Forest Economist, two clerks and a collector, to which was added in 1910 an Imperial Assistant and later, an advisory officer on questions of paper pulp. It was realised as long ago as 1912 that special sections would have to be formed for each subject of Forest Utilization but it was not until 1920 that anything like a full staff was appointed. The sanctioned staff consists of a Forest Economist and an Assistant Forest Economist, eight officers in charge of Sections with their assistants and subordinates, a sub-section of mechanical engineering and the clerical and miscellaneous or technical staff. The personnel as at present entertained is shown on the accompanying table on pages 6 and 7.

As time goes on there can be no doubt that the staff will have to be still further increased, especially in the Section of Minor Forest Products in which several specialists are urgently required.

III.—The Capital Cost of the New Laboratories and Workshops and the Current Expenses.

The capital cost of the new laboratories and workshops can best be shown under three heads, namely cost of land, cost of buildings and cost of plant, including erection. It is difficult to give accurate figures, for the reason that of the 1,300 acres of land acquired for the whole Institute, including Research, Education and a plot of 300 acres put aside for the proposed Chemical Department, only 92 acres are at present taken up by the Economic Branch. Then again the water, electrical and drainage schemes are common to all Branches and to allot the proportions pertaining to the Economic Branch is by no means an easy matter. With these provisos the figures given for capital expenditure are as follows:—

	Rs.
Cost of land (92 acres)	88,700
,, laboratories, workshops, godowns, log-pond, sawmill and offices	6,00,350
,, all experimental machinery and running plant and erection of same	5,72,700
,, gantries	25,500
,, power house	37,500
,, water supply	2,00,000
,, roads	47,500
TOTAL	15,71,750

The present annual sanctioned budget for the Economic Branch comes to Rs. 3,83,000, that of future years when all Sections are fully functioning will be in the neighbourhood of Rs. 4,50,000.

IV.—Aims and Functions of the Economic Branch.

Well may it be asked to what this large expenditure on plant and staff will ultimately lead. What are the aims and functions of the Economic Branch ? And finally, what has been done in the past to justify the existence of such an organisation ?

To answer the above questions it is first necessary briefly to review the position of the market with respect to Indian timbers. We have in all some 2,500 or more species of trees in the forests, not to mention another 2,500 woody shrubs. Of those species which grow to commercial sizes, over 500 are available in commercial quantities. Again, of these 500 or more species, the annual outturn of about 60 can be sold up to their available limit, of another 160 species the supply is in excess of the demand and for the balance there is little or no demand. The same is the case with minor forest products, some of which can be disposed of to the full amount available, others can only partially be disposed of, while many others are of great potential value, but until the best methods of preparing them and placing them on the market have been discovered, they are unsaleable.

Having briefly reviewed the position of affairs it requires little imagination to understand why extensive research work is necessary. However, it is necessary to state clearly what are the aims and objects of so large a Research Institute, otherwise continuity of effort and coordination of work must necessarily suffer. The aims and functions of the Economic Branch may be summarised as follows :—

(1) To find markets for timbers and minor forest products the supply of which is in excess of the demand.

(2) To act as an advisory agency to users of timber and minor forest products.

(3) To indicate to enquirers the localities where supplies may be obtained.

(4) To advise, when requested to do so, on schemes for the development and organisation relating to Provincial problems.

(5) To act as a liaison agency between users of timber and the local Forest Officers, and especially of administrative units dealing with utilisation.

(6) To collect statistics of all kinds referring to current prices, supplies, market fluctuations, imports, etc., and to issue periodical Trade Notes on the same.

The work to be carried out under the first heading often involves much research work. Anyone who has been confronted with the problem of placing a new timber on the market will well understand the difficulty of doing so. The prospective user is invariably suspicious, he will find one hundred and one faults with the timber, though he cheerfully uses a similar timber which develops far more eccentricities, such as warping, splitting, contracting, discolouration or even decay. Then again, the timber to be brought on the market may be difficult to season or it may not be sufficiently durable and may require special treatment, or it may " wind " when coming off the saw and may require to be cut in some special way, all of which and many other problems require close investigation. If the timber is considered to be best marketable as beams and rafters, or for some special purpose such as oil-well sucker-rods, gun carriage shafts, hammer handles etc., a long and special series of timber tests will have to be carried out before the people interested in such articles are convinced that the timber is sufficiently strong. Then again, in the case of minor forest products, such as essential oils, drugs, turpentines, gums, resins and oleo-resins, the experiments may lead to tapping of trees to determine the best method of collection in order to obtain the greatest yield, or chemical analyses may have to be carried out in the laboratory, followed by semi-commercial tests. In the case of raw material for paper pulp, in which India abounds, both laboratory and semi-commercial tests are required before any papermaker will even consider the question of carrying out preliminary investigations of his own with the object of utilizing the material. All the work under the other headings can only be satisfactorily carried out by collating, tabulating, and recording the information available from the experiments carried out under the first heading and from past records.

Before leaving this subject it is interesting briefly to review the position of Forest Economic Research in India, as compared with that in other countries. In Canada and America the Forest Product Laboratories, which are a counterpart of the Economic Branch in India, have behind them a large number of wood-working institutes, controlled by keen business men, who are only too ready to make use of and co-operate with the Research Officers. In India we have few wood-working industries and what there are, are scattered and often far from the Forest Institute It has therefore been found necessary for the provinces to start Utilization Circles, or organisations to carry on the early stages of development, leading to more intensive utilisation of their timbers,

Photo by T. B. Chitrakar.

Forest Economist's Workshop, 1913-1923.
Old Forest Research Institute, Dehra Dun.

[To face page 10].

Photo by Nathu Ram.

Creosoting Plant at Dhilwan, N. W. Railway. Treated sleepers being removed from cylinder.

[To face page 11].

and to translate the results, obtained by the Central Institute, into commercial propositions.

V.—Justification for the Economic Branch.

The capital cost of land, plant, and machinery at the new laboratories and workshops of the Economic Branch, as given on page 8, amounted to Rs. 15,71,750, while the annual current expenditure amounts at present to Rs. 3,83,000 and will probably increase to Rs. 4,50,000 in future years. These are large sums of money, which require full justification.

To give hard figures in rupees and annas of revenue derived from investigations carried out by the Economic Branch is not an easy matter for the reason that the Forest or other Departments, as well as commercial concerns do not always report the results obtained from information supplied. Then again, it can readily be understood that in commercial undertakings it is often inexpedient to publish financial results. On the other hand it is possible to state, in broad terms, the results of certain experiments which will show that good interest is being obtained on capital and current expenditure. As an instance in point, the North-Western Railway, as the result of the experiments started in 1910 with antiseptically treated sleepers, put up a large creosoting plant last year, capable of treating some 400,000 B. G. sleepers annually. The timbers treated by them were never before used as sleeper woods and the department had difficulty in finding a market for these species. The financial gain on these transactions alone would pay a considerably higher rate of interest on the capital expenditure of the Economic Branch than many thriving commercial concerns. A similar plant dealing with from 250,000 to 300,000 M. G. sleepers in Assam has been in operation for more than a year, and this plant was based on experiments carried out by the Forest Economist at Digboi in 1915. Another instance is that of paper pulp which, after prolonged investigation, has led to the erection of a pulp factory, which works with bamboo obtained from the State forests of the Chittagong Hill Tracts. Yet another instance is that of rifle-stock wood, which until the Research Officers went into the question, was all imported from America and which is now entirely supplied from the North West Frontier Province and Kashmir. Numbers of similar instances can be mentioned such as the turpentine and resin industry, the rosha-oil industry, match, pencil and penholder factories and Indian timbers used for bobbins, hammer-handles, billiard cues, railway wagons, sporting requisites, ply woods, etc., all of which were not formerly utilized for such purposes and of which the supply was in excess of the demand. A perusal of the account of work carried out by each section of the Economic Branch should remove all lingering doubts that may exist in the minds of interested persons as to the justification for the existence of the Economic Branch. If further doubt exists, the public is cordially invited to visit the workshops and laboratories at Dehra Dun.

VI.—Miscellaneous Facts of General Interest.

Not only do the museums of the Forest Research Institute attract large numbers of visitors both on business and on pleasure, but the laboratories and workshops are equally popular. The museums are open to the public, who take ample advantage of the privilege, while many people also take the trouble to go out to Kowlagarh to visit the laboratories and workshops. Recently two parties of members of the Legislative Assembly paid a visit to the Institute, and from remarks made it was evident that they were extremely interested in what they saw. Many distinguished visitors have visited the museums and works, including His Excellency the Viceroy, the Commander-in-Chief, the late Governor of the Punjab, the Finance Member of the Viceroy's Council, the Chief Commissioner of Railways, the Conference of Chief Engineers of Railways in India, the Tariff Board, senior officers of the Ordnance Department, many distinguished Generals and naturally many Forest Officers and business men, and representatives of foreign countries.

When the Economic Branch first started, one solitary enquiry, and that of no great importance, represented the daily post and perhaps that is an overestimate, while a liberal estimate may put important enquiries received at that time at 20 a year. How many are now received it is difficult to estimate, as a large number are still of little importance, though very numerous. A conservative estimate of important letters dealt with, requiring detailed reports and involving often months of research work and investigation, may be put at 600 per annum. This is but a fraction of the work, the real results of long investigations, dealt with according to carefully prepared plans of operation known as "Projects", are issued as Bulletins, Records and Memoirs of the Economic Branch by the Forest Research Institute.

A Trade Supplement to the "Indian Forester" is issued monthly, giving prices current, all over India and Burma, of timber and minor forest products. Short trade notes are also added and notices of sales and auctions. This Trade Supplement is much appreciated by timber dealers and the local Forest Officers, and as staff and funds permit, it is hoped to enlarge this publication.

Spare collections of hand specimens of timber are issued free of charge to *bonâ-fide* enquirers, while a representative exhibit of specimen planks, collected from all parts of India, was recently sent Home for the British Empire Exhibition.

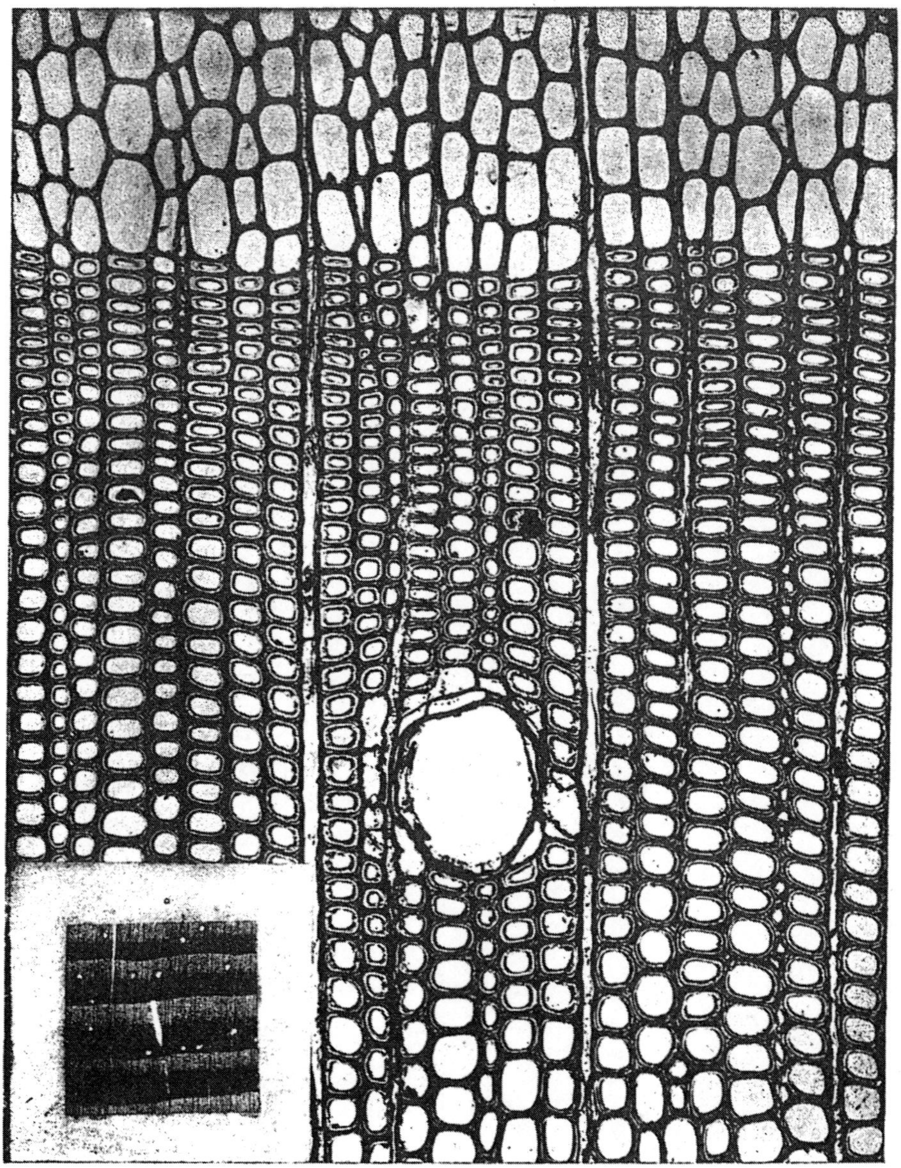

A specimen microphotograph and macrophotograph (inset) of *Pinus Merkusii.*

[To face page 13].

II.

Section of Wood Technology.

The Section of Wood Technology may be said to be the basis of all timber utilisation, as it deals with the structure of the timber, that is, the elements of which wood is constituted and the arrangement of the same, on which the quality of any individual timber depends. The term "Wood Technologist" is somewhat vaguely used, as it covers several branches of the work. It may apply to an expert in the working qualities and uses of timbers, or again, an expert in either seasoning or timber testing may justly claim to be a Wood Technologist, but as far as this Section of the Economic Branch is concerned, by Wood Technology is meant the study of the anatomical structure of timber and the application of knowledge in this subject to practical propositions.

To start the Section it was fortunately possible to obtain the services of Dr. H. P. Brown, an eminent scientist and Professor of Wood Technology, State College of Forestry, Syracuse University. He was lent to us by that University for 22 months, and during that period he prepared photomicrographs of over 180 of our more important timbers, organised the Section, and prepared a most valuable Elementary Manual on Wood Technology in terms of Indian Timbers for instructional purposes at Forest Colleges in India. (See illustration of a photomicrograph on opposite page.)

Dr. Brown has now returned to America but the Government of India have been able to retain his services during his vacation periods during which time he will continue the preparation of microphotographic sections, prepare a manual on Indian timbers in collaboration with Mr. Pearson, the Forest Economist, and train an officer in Wood Technology for this Institute.

It may be of interest to describe briefly the activities of this Section of the Economic Branch. Primarily the work consists in the study of the structure of Indian timbers, the upkeep and enlargement of the wood collections, and the preparation of micro and macrophotographic collections. The routine work chiefly entails the identification of timber specimens sent by forest officers, wood working institutes, railway officials, firms and others interested in timber propositions.

It may well be asked what is the practical use of the critical examination of the anatomical structure of our timbers? The answer is threefold. Without a definite knowledge of the structure of a timber it cannot be definitely identified, and in this connection the microphotographs are of the greatest assistance. As an example of the value of such work, a large municipality had contracted for some buildings and specified the use of certain timbers. They suspected that the correct species of timbers were not being employed and so sent samples to the Institute for identification with the result that the contractors were severely punished. Then again, one of our most difficult problems is to find new uses and markets for many of our timbers. With a full knowledge of the anatomical structure of Indian timbers it is possible to select those timbers which closely resemble the timber already in use for a specific purpose for which a substitute is being sought. As an instance in point the selection of Indian timbers to take the place of imported ash and hickory for hammer handles, shafts, etc., may be cited, and has been achieved with considerable success.

The third case in point is the preparation of both field and laboratory keys for the identification of Indian timbers which can only be accomplished after a long and careful study of the subject.

Enough has been said to demonstrate the great importance of this Section, for which a sound foundation has fortunately been laid.

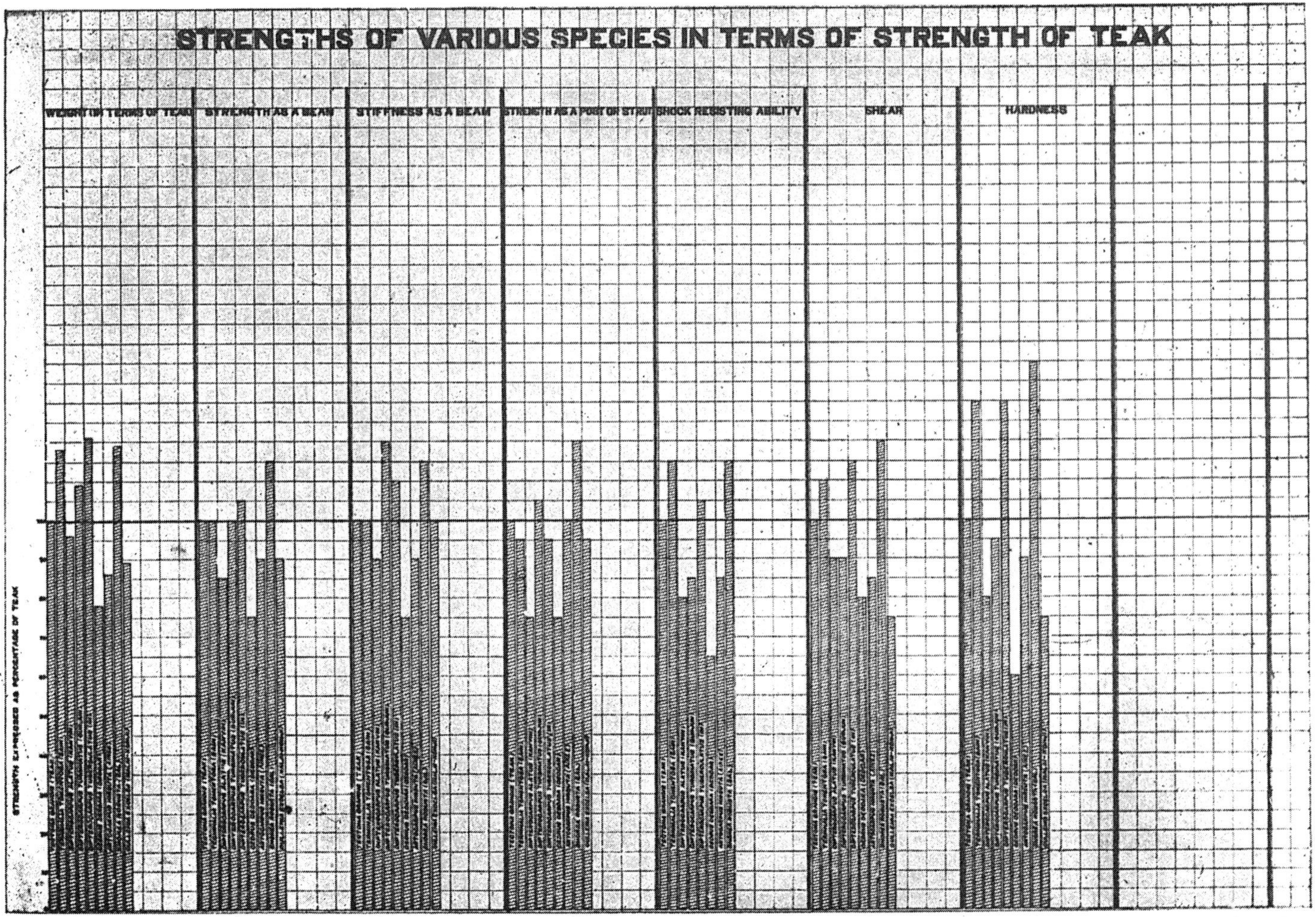

[To face page 15].

III.

Section of Timber Testing.

BY

L. N. SEAMAN, *Officer in Charge, Timber Testing.*

One of the most important groups of investigations conducted at the Forest Research Institute, both from the point of view of quantity of work done and of practical value of results obtained, is that which relates to the investigation of the mechanical properties of Indian timbers. Though this branch of research is still in its infancy, over fifteen important species have been fully tested and many others are being dealt with at the present time. Ten testing machines are kept in constant operation, and two more are to be added to the equipment as soon as possible. In spite of this fact the need for reliable information on the strength of Indian timbers is so great, and important enquiries from timber using industries, railways, and government departments so numerous that the Testing Section has far more work than it can cope with for many years, and only the most urgent problems can receive immediate attention and prompt solution. To meet this great and growing demand the staff are working at utmost capacity and turning out from 1,000 to 1,500 strength tests and a corresponding number of physical determinations every month.

A very little thought will show the vast economic value of this class of research work. India possesses a wealth of valuable timbers which few other countries can approach, and yet, because of lack of information as to their strength characteristics and fitness for certain purposes, imported timbers, very inferior to what might be obtained locally, are used. A few of the standard Indian timbers, it is true, are well known, but scores of excellent varieties are known as jungle-woods, and valued only as fuel, or not even as that. The results obtained by the Timber Testing Section are daily demonstrating new uses and new values for these woods which up to the present have been mere forest weeds.

The main basis of this research lies in a programme of routine tests conducted by standard methods, and therefore suited both to establish the relative strengths of well known and unknown Indian woods, and also their relative strengths as compared with well known foreign species of timber. From this work is gradually being built up a mass of reliable data, constantly growing in value, and serving as a starting point for all the special investigations necessary to determine the suitability of Indian woods for specific uses. Tables and charts, an example of one of which is shown on the opposite page, are prepared from these data showing the relative usefulness of different woods for beams, posts.

sleepers, scantlings, bridge parts, and other common uses. These data combined with others obtained by testing full sized structural timbers form the basis for reasonable grading rules, based on the strength of the timber and for preparing tables of allowable working stresses for use in the design of bridges, buildings, and other structures. It is evident that, if by intelligent grading and good design even a little timber can be saved on each structure, the net annual saving to the country will be very great indeed. When we add to this the fact that the same structures can be built of what are at present unknown, or little known Indian timbers, and better so built than of expensive imported timbers, we have not only a saving to the builder, but a gain to the country, in that what the builder has spent, he has spent in India.

Suggestions for such grading rules and tables of working stresses have already been prepared and submitted to the Military Works Department, the Punjab Public Works Department, and other important concerns. Their advantages include safer structures and more economical use of timbers on account of the definite data furnished for the architect and the engineer, and the possibility of using the lower grades of timber for purposes such as wall-joists and small buildings where the stiffness rather than the strength of the timber is the principal consideration. They furnish a definite guide for the placement of each quality of timber in the situation for which it is best suited. Tentative grading rules for *chir* pine (*Pinus longifolia*) railway sleepers have recently been drawn up, and much similar work is under consideration.

These two main lines of testing, namely the routine tests of small clear specimens, and the tests of full sized structural timbers are supplemented by many other lines of investigation, each having for its object the selection of Indian timbers for some specific use.

At the present time a very extensive study is being made of all Indian timbers likely to prove satisfactory for use as railway sleepers, including tests of spike-holding power, and lists are being prepared showing which woods make the very best sleepers, and how these woods compare with foreign sleeper woods. As the same kind of work is being done in the United States Timber Testing Laboratory, and lists prepared in the same way, it is interesting to note that if creosoted Indian *sain* (*Terminalia tomentosa*) be placed in the American list it takes *second place*, being surpassed only by black locust (*Robinia pseudacacia*). It stands well above white oak (*Quercus alba*), one of the most highly valued sleeper woods in America, and the figure expressing its sleeper value (in quality, but not in price) is more than double that figure for Douglas fir (*Pseudotsuga mucronata*). The results of the sleeper study alone, in increased use of Indian woods and decreased use of imported woods, will repay to India many times over the entire capital cost and running expenses of its timber testing laboratory.

Photo by I. N. Sharma.

Timber Testing Workshop showing impact machine, universal 30,000 and 50,000 lb. testing machines.
New Forest Research Institute, Dehra Dun, 1924.

[To face page 16].

Another interesting investigation is that which relates to the discovery of Indian and Burmese substitutes for American hickory as oil-well sucker rods. When the routine tests indicate that a species of timber possesses the qualities that are necessary for this purpose, actual sucker rod ends are made up of that timber and subjected, in the testing machine, to as nearly as possible the same stresses as they would have to endure in the oil well. Although a number of kinds of timber have failed in this trial, four have already been found as good as, or better than hickory for the purpose, and two of these, *sundri* (*Heritiera minor*) and *sain* (*Terminalia tomentosa*), are already being made up and given a final trial in actual use in oil wells in Burma. Up to the present no wood except American hickory has been used in India and Burma for this purpose, and in one year a single company imports about eight lakhs of rupees worth of hickory sucker rods. The results of this study will mean the cessation of this import and the spending of the money within the country, as well as the employment of Indian labour in the manufacture of a better article than the imported rod.

Then there is the study relating to material for gun carriage poles and other ordnance and vehicle parts undertaken at the suggestion of the Superintendent of the Gun Carriage Factory at Jubbulpore. This work has scarcely started as yet, but will have for its object an analysis of existing designs, and the possibility of new designs, and of substituting timbers at present not recognised as suitable for the purpose.

A study of boxes, especially tea boxes, has recently been started at the Testing Laboratory. These, like all other tests in the Research Institute are conducted along the most practical lines, the object being to supply information directly useful to producers and purchasers. Standard boxes, recognized as satisfactory, are first made up in the usual way and tested, and the results of these tests taken as the basis of comparison for proposed substitute woods or changes in design. Some boxes are so placed in the testing machine as to be crushed by pressing on opposite edges, and others are dropped successively on their corners till they break and spill their contents. Later, it is proposed to add a drum box-testing machine, a standard device by means of which a box can be subjected in a short time to almost the same usage that it would have to withstand in a long journey. This machine will give a very accurate comparison of the quality of different boxes tested in it.

Though standard boxes are taken as the basis of comparison in this investigation it is not proposed to assume that they are the best that can be produced. The results of the tests on these boxes will be analysed and studied to determine whether it is not possible either to produce an equally good box at a lower price, or a better box at the same price.

Somewhat closely related to each other are the tests of picker arms and of axe and hammer handles, as both involve a considerable amount of impact testing. The former is a new study, but the latter has been in progress for some years and has produced good results. As in most other cases the suitability of a wood is first judged from the results of routine tests, and then actual handles are made up and tested as such. Results of this work have been published from time to time. As a direct result of the report issued recently on tests of *Parrotia Jacquemontiana*, the Military Stores Department has decided to place a contract for handles of this material, and in this way another "forest weed" is coming into its own.

This is a very brief sketch of a little of the work that is being done by the Timber Testing Laboratory, but it is sufficient to show the value of the research which is demonstrating the worth of many of the so-called jungle woods, frequently proving them to be superior to expensive imported varieties, and showing how various timbers can be used to the best advantage. No research institution is ever meant to be a direct profit producing concern, but in an indirect way it brings increased revenue to the country that supports it. From the few instances cited above it is evident that this Laboratory like every other Section of the Forest Research Institute, is the cause of large contributions to the country's revenue, and considered even from the financial point of view it is an extremely good investment.

Photo by Har Swarup.

Timber Testing Workshop, showing 18 ft. constructional beam being tested.
New Forest Research Institute, Dehra Dun, 1924.

[To face page 18].

IV.

Timber Seasoning Section

BY

S. FITZGERALD, *Officer in Charge, Seasoning.*

Although the difference between seasoned and unseasoned timber is only too obvious to the ultimate user of the timber, few manufacturers will take adequate care to use only properly seasoned material in their products. Some of the more responsible keep stocks for a considerable period, trusting that it will be ready when required for use, but few employ any definite means of proving the condition before use. Others buy supposedly seasoned stock (which generally is not seasoned) and the general result, we all know from bitter experience, is that furniture and joinery shrink and generally give trouble.

So much woodwork is spoilt through unseasoned wood being used, so much good timber is lost through improper treatment in seasoning, that the Seasoning Section has a very large task to perform, and one which will materially assist Indian woodworkers and benefit eventual users.

The work of this Section is first to find out how best to do these things under Indian conditions, and then to hand on the information, showing people how to judge degrees of seasoning by simple tests, advising as to the care of timber stocks, seasoning under open air conditions, and discovering and communicating the best processes of accelerating the drying by mechanical means so as to decrease the wastage of material and interest on capital, and to assure a properly seasoned product.

Open air seasoning trials were started in 1914 by the Forest Economist, collaborating with the Chief Conservators and Conservators of Forests of India and Burma. Ten thousand planks, scantlings and sleepers, were put under observation. These were from 33 species of Indian timbers and they were arranged in 22 different stations with widely varying climatic conditions. Each experiment, with each species and at each station, involved seven different ways of manipulation. Some trees were girdled before felling, some sawn green, others water-immersed for a period before sawing, some barked, etc.

The results of these experiments are given in detail in the Indian Forest Record, Volume VII, Part I, but summarised, the general result was to show that it profited to girdle some species and not others, that certain timbers parted with their moisture more easily if water immersed for a period, and that others were best sawn up green directly after felling. Above all, the importance of careful stacking and protection from sun and weather, were clearly proved.

Experience gained in these first trials made it possible to eliminate some variants in a subsequent experiment and to concentrate more on ascertaining the value of girdling, water-seasoning, and the period to allow between felling and conversion.

In the second series of open air seasoning experiments, 53 species of Indian timbers were under observation, and as before, in different parts of the country.

The amount of work and travelling about, supervising these experiments being considerable, a Seasoning Officer (a specialist from the Forest Products Laboratory, Madison, United States of America), was appointed to continue the work, which was until this time done by the Forest Economist with the co-operation of the local Forest Officers. The full results of the second series of open air seasoning experiments are recorded in Indian Forest Record, Volume IX, Part V.

Further open air seasoning experiments are now being completed on Andaman timbers, and also on a few Indian timbers for special purposes.

Another series of open air seasoning tests will be made with sleepers, following their course from the time of felling until delivered eventually to the Railway Company. This will be detailed elsewhere in Project IV of the Economic Branch.

Whilst open air seasoning work was taking its course, preparations were proceeding for experiments with "artificial" seasoning processes. Firstly, a roughly constructed timber dryer was set up in the Institute grounds at Dehra, (see photograph on opposite page) and later, at Kowlagarh site, fully equipped Timber Dryers on the two leading systems :—Tiemann System—with circulation by water sprays (as used mainly by the United States Forest Laboratory) and Sturtevant System—having blower circulation (as used by the British Government for aircraft timbers) were erected under the expert advice of the Seasoning Officer.

Description of Apparatus.

The Tiemann outfit consists of three units each, 24ft.×8ft.×10ft.

The heating is by means of steam pipe radiators in the basement of the compartment, and the circulation by thermal means accelerated by water and steam sprayers in a channel at one side of the dryer. The water sprayers are fed by steam driven pumps.

The Sturtevant outfit consists of 4 small compartments, 15′×8′×7′, heated by external steam pipe radiator and the circulation is actuated

Photo by T. B. Chitrakar.

Temporary Timber Seasoning Kiln.

Old Forest Research Institute, Dehra Dun, 1922.

[To face page 20].

Photo by I. N. Sharma.

Sturtevant Kiln (open) showing timber-testing specimens and planks stacked ready for drying.
New Forest Research Institute, Dehra Dun, 1924.

[To face page 2].

by a duplex blower, one part of which recirculates partly used air. the other part, fresh air. The duplex blower is driven by an electric motor. The controls are so arranged that two small compartments can be used as one long chamber to take 30 feet long timber. (See photographs.)

In all cases automatic recording thermometers and hygrometers keep careful record of all the happenings.

Even the first roughly constructed dryer showed the importance of artificial seasoning for Indian hardwoods, and it was proved that timber for the most exacting Ordnance work could be brought to maturity in a few weeks instead of years, and with less degrade than usual. As a result, an important Ordnance Factory is installing dryers which will enable it to reduce considerably the capital locked up in holding timber stocks, and permit a big increase in output in the event of any emergency.

The departure of the first Seasoning Officer (for family reasons) caused a hiatus in the work and some months elapsed before a new man (the officer previously in charge of seasoning for the Air Ministry) could be obtained. However, in the few months during which the new seasoning plant has been at work, 12 species of Indian timbers have been artificially seasoned successfully in bulk, and about 30 others tried in a tentative manner.

Most of these have reached complete seasoning with less deterioration than is commonly allowed for open air seasoning, and in one case a timber, available in large quantities, but commonly regarded as worthless because of its susceptibility to fungus attack and bad discolouration, has by artificial seasoning been brought out as a good packing case timber.

It will take a long time to establish the best treatment for the many valuable Indian timbers, but eventually the drying specification for each of these and for varying thicknesses of each, will be available in a form that can be used by commercial firms who provide themselves with suitable apparatus.

It may occur to some to question why this seasoning business has suddenly become acute. The reason is that the supplies of the well-known good timbers such as teak, deodar, and *sal*, which could be easily seasoned and worked, have lately proved insufficient.
Greater demand and shorter supplies caused prices on these few species to go up, and other timbers had to be brought into use.
These many new timbers have to be studied and the best treatment discovered, so as to put them into use in the most attractive manner.

The Seasoning Officer—a specialist with 18 years' practical experience of timber seasoning—is at the disposal of all who are using Indian timbers, to study their troubles and devise means of getting over them, to advise as to best type of dryer and give specifications of apparatus, etc., and to help with the working when installed.

The Seasoning Section will train operators in the working of seasoning apparatus, and, if desired, men can be sent for a short course of training with the Institute seasoning kilns.

Already there is considerable demand for help, and schemes have been worked out *gratis* for enquirers in Hyderabad, Madras, Burma, Calcutta, Bareilly and Lahore, and in some cases trial loads of firms' own timbers have been seasoned at Dehra, to show the result of artificial seasoning for their own particular purposes.

The timber dryers at Dehra have all the elaborations required for research work and the necessary instruments for keeping exact records of processes attempted and results obtained, but at the same time the compartments are of commercial size, taking from 2 to 8 tons of timber at a charge.

Small scale experiments may lead to disappointments when expanded. The artificial seasoning research at Dehra is done on a commercial scale, and the results can be communicated to, and repeated by, a commercial firm.

Photo by I. N. Sharma.

Control Room, Sturtevant Kilns,
New Forest Research Institute, Dehra Dun, 1924.

[To face page 22].

V.

Wood Preservation Section

BY

J. H. WARR, *Officer in Charge, Wood Preservation.*

HISTORY.

The history of this Section is practically the history of the antiseptic treatment of timber in India, for although earlier investigations into the possibility and desirability of preserving Indian timbers from decay and destruction by insect pests had been undertaken, no conclusions had been arrived at, and no recommendations made, and it was left for the present Forest Economist to institute a series of durability tests on a number of different species of timber treated with preservatives and also on the preservative value of different antiseptics.

As little or nothing was known in India, previous to these experiments, about the antiseptic treatment of timbers, and owing to the complete absence of results from former investigations conducted many years earlier, it is not to be wondered at that the attitude of railway engineers was one of caution and rather sceptical interest.

An almost total lack of funds was a very difficult obstacle to surmount, more so as want of interest precluded any idea of obtaining financial support for the purchase of an elaborate experimental plant. The first attempts were therefore made with the simplest and cheapest apparatus obtainable, namely, the "Open Tank"; that is to say the timbers to be treated were immersed in hot oil contained in open topped tanks for varying periods and then allowed to cool so as to absorb the antiseptic.

The original scheme for the "Open Tank" experiments was laid down in 1910 and brought into effect at once, while laboratory experiments on the value of different antiseptics were started in January 1909.

The enormous value of these early experiments is becoming more evident daily, especially as, for various reasons, the interest in the artificial preservation of timber becomes more stimulated.

Briefly, the laboratory experiments consisted in treating a dozen different species of timber under strictly comparative conditions with no less than 23 different antiseptic salts and oils and placing them in the ground as stakes, together with stakes of untreated timber of the same dimensions. These untreated specimens acted as controls.

A complete set of comparative data concerning the effectiveness of different antiseptics both against decay and insect attack is now in existence, and provides valuable information for those who may wish to adopt them for use. A diagram showing the comparative protection given by various antiseptics on the same species is shown opposite page 27. The open tank experiments were carried out almost exclusively on possible sleeper woods, which, after treatment, were laid in various sections of the railways all over India and kept under observation. Although the number of species was necessarily restricted the experiments conclusively proved the value of the artificial preservation of inferior woods ; so much so that at least two large and one small preservation plants have been erected and are now in operation, while a number more are contemplated.

The service records of the sleepers treated in the above experiments are published annually, the latest publication being Forest Bulletin 59 of 1925.

It must also be remembered that the places where the service tests on the treated sleepers were carried out were carefully chosen with a view to ascertaining what influence, if any, the varying climatic conditions of these localities had upon the effectiveness of the different antiseptics used and upon the treatment to which they were subjected. It can at once be stated that the locality has little influence upon the results and that proper antiseptic treatment is effective all over India. As explained above, the absence of an experimental pressure plant in India restricted somewhat the scope of the investigation, but, in 1915 experiments on a few Assam species were carried out at Digboi in a pressure cylinder, the property of Messrs. The Assam Oil Co., Ltd., Digboi, Assam, with very encouraging results.

It is quite evident from these experiments and the service records of the sleepers therein treated, that a number of timbers hitherto regarded as inferior and practically valueless can be brought into useful service and are thereby considerably enhanced in value. In fact, it is not too much to say that some timbers have actually been given a value by antiseptic treatment as formerly they were either not extracted or if extracted were only used for fire-wood.

Meanwhile the interest of the Railways, aroused by these experiments, had been further stimulated by the advance in price due to the increasing scarcity of the naturally durable sleeper-woods such as *sal*, deodar and *pyinkado* and it was decided to have a really up to date pressure plant erected at Dehra Dun for the purpose of carrying out further experiments on an increased number of species and to increase the staff at the Institute. The war, however, delayed the carrying out of these schemes and it was not until 1922 that further steps could be taken. The plant decided upon consists essentially of a cylinder 3 feet 6 inches in diameter and about 11 feet long, surmounted by a small

Photo by I N Sharma.

Creosoting Plant of Assam Railways and Trading Co., Ltd., Marghuerita, Assam. December 1923.

[To face page 24].

Photo by I. N. Sharma.

Experimental Wood Preservation Plant, showing pressure cylinder (open), storage tanks, etc.

New Forest Research Institute, Dehra Dun, 1924.

[To face page 25].

dome and capable of withstanding a working pressure of 200 lbs. per square inch. It is closed by a door fastened with screw clamps.

Underneath the cylinder is a square service tank of about 800 gallons capacity fitted with steam heating coils and a thermometer which registers the temperature of the contents at any particular moment. It is now fitted with a pneumercator gauge which indicates in pounds the weight of the liquid in the tank.

In addition, at the back of the cylinder and on the floor level, are two cylindrical storage tanks of about 500 gallons capacity used for storing the various antiseptics to be used. An overhead pressure tank of about 600 gallons capacity has now been added for the purpose of working what is known as the Rueping Process. (See photograph opposite.)

Two pumps with temperature and pressure gauges on the pressure cylinder and over-head pressure tank complete the equipment. All these gauges are mounted on a board in a convenient place facing the operator.

The two pumps are worked from the same shaft by an electric motor and consist of plunger pumps, one for pumping oil into the cylinder and also for raising the pressure, and the other to act as a vacuum pump and an air pressure pump. The vacuum pump is connected to the dome of the cylinder through a receiving vessel and a water condenser to prevent any liquid getting into the pump cylinder and thus causing damage.

The treating, or pressure cylinder, is capable of taking 15 broad gauge or 24 metre gauge sleepers at a time, and is fitted with a shelf on which the sleepers under experiment rest, steam heating coils, vacuum and pressure gauges, and also a registering thermometer.

The conditions under which the sleepers are being treated with respect to pressure, temperature and vacuum are thus known and can be controlled by the operator.

The *modus operandi* of the plant is as follows. Previously weighed and measured timber is put in position in the treating cylinder and the door tightly clamped by means of the screw clamps.

The wood may then be subjected to a vacuum or not, according to the nature of the timber being treated. Oil is then sucked from the underground service tank into the cylinder, and when this is full, all valves and cocks are closed with the exception of those from the pressure pump. More oil is then pumped in until the requisite amount has been absorbed by the wood. The oil is then released and returned to the service tank. After the wood has been subjected to a vacuum of varying intensity and duration according to circumstances the treatment is complete and the timber removed.

The erection of the plant at the new site of the Institute was completed in 1923 and a few experiments were carried out that year on spruce, fir and *Terminalia tomentosa*.

It was then decided to obtain the services of an expert in wood preservation and increase the staff, so that the work is now under the direction of a Section Officer assisted by an Imperial Assistant.

In 1923 a scheme was drawn up known as "Project IV, Mechanical Strength, Seasoning Properties, Treatment of and Key to, certain Indian Sleeper-woods."

The object of this Project was to correlate all the known or determinable factors which underly the choice of a particular timber for railway sleepers, based on the assumption that to ensure durability the timber requires treatment. This scheme has been submitted to all the railways of India and enlarged and approved by them.

At present the number of species which will be experimented upon exceeds sixty and is being added to as the scheme becomes known and as the requirements of different localities become more apparent.

The work of the Wood Preservation Section in this Project is chiefly to determine to what extent a timber can be impregnated by various antiseptics, to determine its durability when so treated, and to draw up specifications for its preservation, so as to obtain the maximum economic value therefrom.

Two hundred sleepers of each species will be impregnated, if amenable to treatment, with various antiseptics, and records of service, on the lines of those previously made, will be kept for the purpose of ascertaining the value and effectiveness of the process.

Previous experiments have shown that if a suitable antiseptic is used and a certain depth of penetration obtained the antiseptic treatment of timber is as effective in this country as in many others, but the *economic limits* have not yet been determined for various species and different localities, and this is the main object of the present enquiry.

For instance, it may be possible to inject so much antiseptic that the timber is broken up by wear long before it fails from decay, or too little may be absorbed so that it decays before it becomes unserviceable from mechanical defects. A wastage either of antiseptic or of timber is thereby caused, and it is to determine the treatment required so that the timber fails from both causes at the same time that this research work has been undertaken, for, obviously, this is the most advantageous treatment in the long run.

Although the principal work of the section is on railway sleepers, other departments of constructional work are by no means overlooked, for the arguments which apply to the need for preserving railway sleepers from decay and destruction, apply with equal force to the timbers used in building construction, street trades, paving, telegraph and telephone and many other technical trades.

Photo by I. N. Sharma.

Timber Godown, showing sleepers stacked for air-drying prior to treatment in wood preservation plant.

New Forest Research Institute, Dehra Dun, 1924.

[To face page 26].

ABSTRACT OF THE ANTISEPTIC GRAVEYARD RESULTS
FOREST RESEARCH INSTITUTE
DEHRA DUN.

No.	Antiseptic.	Condition.	Boswellia serrata. 1	Pinus longifolia. 2	Pinus excelsa. 3	Picea Morinda. 4	Abies Pindrow. 5	Pterocarpus macrocarpus. 6	Bombax malabaricum. 7	Bauhinia retusa. 8	Dipterocarpus tuberculatus. 9	Anogeissus latifolia. 10	Odina Wodier. 11	Shorea robusta. 12
1	Avenarius carbolineum	Treated (T)	3	14_d	10	14	4	C_m	4	6_m	6_m	6_m	3	6_m
		Untreated (U)	1	14_d	2	1	2	6_m	1	3	5	3	2	3
2	Jodelite	T	4	14_d	11	3	4	14_d	3	5	14_d	14_m	3	14_d
		U	1	14_d	3	1	1	14_d	1	3	14_d	3	1	2
3	Sideroleum	T	8_s	8_s	8_s	8_s	8_s	8_s	8_s	8_s	8_s	8_s	8_d	8_s
		U	1	5_d	7	1	2	7	2	2	6	1	5	8_s
4	Brunolinum	T	8_s	8_s	8_s	8_s	8_d	8_s	8_s	8_s	8_s	8_s	8_d	8_d
		U	4	5	7	2	3	8_d	1	3	5	1	3	1
5	Green oil	T	9	10	10	4	13	12	11	4	9	11	4	13_s
		U	10	1	2	1	1	8	1	8	4	2	1	13_s
6	Shaw Wallace's creosote	T	6_s	6_s	6_s	6_s	6_s	6_s	6_s	6_s	6_s	6_s	6_s	6_d
		U	2	3	6_d	3	3	6_s	1	2	5	8	3	6_d
7	Cresoyle	T	13_d	10	13_d	13	12	13_s	13_d	6	5	6	1	13_s
		U	1	2	1	1	1	13_s	2	2	9	2	1	13_d
8	Anthrol	T	12_s	12_s	12	12_s	12_s	12_s	12_s	12_s	12_s	12	12_s	12_d
		U	1	1	1	1	1	5	1	1	12_d	2	1	12_d
9	Concentrol	T	6_s	6_s	6_s	6_s	6_d	6_s	6_t	6_d	6_s	6_s	6_s	6_d
		U	1	4	5	2	2	6_d	1	1	4	3	1	6_d
10	Lignolite	T	8_s	8_s	8_s	8_s	8_d	8_s	8_s	8_d	8_s	8_d	8	8_s
		U	1	5	7	5	2	8_d	1	3	7	4	5	8
11	Burma oil	T	3	11_d	8	7	2	11_d	4	8	7	6	7	11_d
		U	2	2	1	2	2	11_d	1	1	6	5	1	11_d
12	Liquid fuel (Borneo)	T	8	11_s	10	5	11_d	11_d	8	7	10	2	3	11_s
		U	1	1	1	1	3	11_s	1	3	7	2	1	11_s
13	Mortant	T	1	2	2	...	3	11_d	11_d	2	10	10	11_d	11_s
		U	1	1	9	...	1	11_d	1	2	1	8	11_d	11
14	Atlas solution	T	3	6_m	5	3	4	6_m	3	4	4	5	3	6_m
		U	1	6_m	3	1	1	6_m	1	3	3	4	1	6_m
15	Bellit	T	2	3	3	2	2	13_s	2	3	11	5	2	13_d
		U	5	1	3	1	1	13_s	1	1	11	2	1	13_d
16	Sodium fluoride & zinc chloride	T	8	9	13_d	3	2	13_d	4	2	9	9	4	13_d
		U	6	1	2	1	2	13_s	3	9	5	2	1	13_d
17	Hylinit	T	5	10	4	3	2	13_d	3	2	13_d	3	3	13
		U	7	3	3	2	2	13_s	1	3	9	3	1	13
18	McDougall's insecticide	T	8	8	8	..	6	11_s	2	3	9	6	9	11_d
		U	1	1	1	...	1	11_d	1	1	2	3	1	11_d
19	Anticide	T	7	12	...	2	1	11	12	...	...	...	...	...
		U	3	2		1	2	1	1	...	...	...	...	...
20	Burnettizine	T	6	1	8	2	6	9_s	1	1	6	6	1	9_s
		U	2	1	8	4	5	9_s	1	2	4	8	1	9_s
21	Cresol calcium	T	1	3	3	1	1	13_d	1	2	8	6	2	13_d
		U	2	5	2	1	1	13_d	1	9	12	3	1	13_s
22	Aczol	T	1	2	7	...	...	10_d	1	8	10	3	2	10_s
		U	1	1	6	..	...	10_d	1	3	8	4	1	10_s
23	Barol	T	9_s	9_s	9_s	9_s	9_s	9_s	9_s	9_s	9_s	9_o	9_s	1_m
		U	1	9	8	4	2	9_d	1	1	9_d	1	2	1_m
24	Pulverized glass	T	1	5	9_c	6	3	9_s	1	...	9_d	4	1	9
		U	1	1	1	3	3	9_d	1	...	9_d	3	1	9
25	Coal tar	T	2	14_d	8	2	3	14_d	3	2	5	5	3	4
		U	1	14_d	3	3	3	14_d	1	2	5	3	1	2
26	Pinus excelsa tar	T	3	2	3	3	3	14_d	3	11	5	5	3	14_d
		U	3	1	1	1	1	14_d	1	10	14_d	5	1	14_d
27	Solignum	T	4	14_d	14_d	11	14_d	14_d	5	7	4	14_d	3	14_d
		U	1	3	18	3	1	14_d	1	3	8	5	1	14_d

LEGEND.—Figures without *any* suffix indicate the life in years.

" with suffix $_m$ indicate specimens missing after $?_m$ years.

" " " $_s$ " " sound " $?_s$ "

" " " $_d$ " " damaged " $?_d$ "

Corrected to November, 1923.

To face page 27.

Fortunately, the results obtained in one set of experiments can be easily translated in terms of other uses so that systematic experiments for other branches of work can for the present be left out in favour of those concerned in the more urgent enquiry on sleepers.

The laboratory experiments started in 1909 on the value of different antiseptics are still being kept up, and, as new preservatives are discovered, so they can be tested on a laboratory scale and placed in their relative class. (See Abstract on opposite page.)

On the more theoretical side an investigation is proceeding to determine the active principle responsible for the prevention of deterioration by fungus and insect attack in the naturally durable timbers such as teak, deodar and cypress, as it is hoped to gain thereby some insight into nature's own methods of preservation.

Past experiments, as previously stated, have shown that *Pinus longifolia* (*chir* pine), previously regarded as a non-durable and therefore unsuitable wood for sleeper purposes, when treated with a suitable quantity of an approved antiseptic is capable of giving at least 12 years' service in the line.

Such an important result could not fail to have its effect and the corollary was the erection of a plant at Dhilwan on the North Western Railway to treat conifers such as *Pinus longifolia* and *Pinus excelsa.*

Other species, such as *Terminalia tomentosa*, the *Dipterocarps*, and some Assam species, have given equally good results and their treatment on a large scale is highly probable. From what has been said it is easy to visualise the outcome of the projected experiments. They aim at establishing a new industry in India which in its turn will have its effect upon the revenues of the country by drawing from the forests those timbers which are at present imported. The large forest tracts in this country are the property of the State and must be regarded as capital invested. In order that this capital may pay interest the forest resources must be developed and utilised, but their value must be amply demonstrated before the consumers are justified in abandoning their present practice or in increasing their future needs.

Such demonstration is obviously the work of the State for, in the first place, no private individual could afford it, and, in the second, the State will be the greatest gainer and, as these experiments have justified themselves in the past, it is safe to say they will more than pay in the future.

VI.

The Paper-Pulp Section

BY

W. RAITT, *Officer in Charge, Paper-Pulp.*

1. The resources in paper-making materials of India and Burma have been known and experimented with spasmodically for over fifty years. Everyone who has touched the subject has been struck by the vast store of such materials growing in the forests, frequently associated with excellent manufacturing and transport facilities, and, in the case of many species, annually or biennially self-reproductive. The introduction to the paper-making world of woodpulp about 1870, however, sent the subject into oblivion for thirty years but, about twenty years ago, interest was revived by the rapid growth of opinion that woodpulp, after all, was not destined to be a final solution of the paper-maker's recurring problem of supplies. The more accessible areas of pulp wood were being rapidly exhausted, costs increased with the opening up of more distant forests and the rapid growth in the world's requirements of constructional timber has created a demand with which the paper-maker cannot compete. The sawmill has become a better market than the pulping plant and the opinion has crystallised that the permanent solution of the question does not lie in a tree which takes 60 to 80 years to grow, which is not easily reproductive and which is of greater value for other purposes. In such circumstances the usefulness of a supply which comes into the market at *waste* value,—there being no other use for it of any importance—and which yields annual or biennial crops, scarcely needs to be emphasised.

2. The present effort to explore such resources began in 1909 when Mr. R. S. Pearson, I.F.S., commenced an examination of several bamboo areas to settle the important points whether the supplies were really so large as rough estimates indicated and whether they existed under conditions favourable for exploitation *in situ* by factories placed in or near the areas of growth. His answer, after prolonged enquiry was emphatically to the effect that there was no room for doubt. The sustained supplies were there in abundance frequently associated with suitable manufacturing and transport conditions and on this reply the systematic experimental work was founded. A commencement was made in a laboratory at the Allahabad Exhibition of 1910-11 under the control of Mr. P. H. Clutterbuck, I.F.S. (now Sir Peter Clutterbuck, Inspector General of Forests to the Government of India) who was Director of the Forestry Section of the Exhibition. When the Exhibition closed, the laboratory, under the writer (now at the

Photo by L. N. Sharma.

Experimental plant in Paper Pulp Hall, showing digesters, beater, mixer and paper making machine.
New Forest Research Institute, Dehra Dun, 1924.

[To face page 29].

Forest Research Institute) was transferred to Dehra Dun. Its scope and usefulness has recently been extended by the erection of an experimental factory (see illustration opposite) on a scale sufficient to permit, and indeed compel, factory methods to be used. The value of such a plant is obvious. It translates the results of laboratory research into practical commercial facts, relieves the industrial pioneer of many pioneering risks, enables him to *see the thing done* and ascertain the principal costs of doing it.

3. The laboratory and field work of the past fourteen years has to a large extent outlined the programme before the factory plant. It has eliminated species which are unsuitable and unprofitable and, with but rare exceptions, has brought about a concentration of effort upon a few families of *Gramineæ*, ranging from small grasses like *sabai* (or *baib*, *bhabar*), to giant grasses like bamboo. With the latter and with savannah grasses, and to a lesser extent with *sabai*, a prejudice has existed to the effect that, notwithstanding the proved fibre value of the pulp they produced, it was difficult to bleach. To the problem of reducing bleaching costs much of the time and effort of the section has been given during the past ten years and, it is satisfactory to be now able to announce complete success. The specific causes of the bleaching trouble have been discovered and located in certain constituents of the complex substances encrusting the fibres and a system of digestion has been evolved which extracts and expels them from the mass before the stage of digestion is reached at which they create difficulty. So long as these discoveries were the results of laboratory work only it was considered inadvisable to publish them until they could be confirmed by trials on a factory scale with digesters specially constructed to facilitate the process. This has now been done and with results which are actually better than those obtained in the laboratory. It is a case where mass tells beneficially. With both bamboo and *sabai* the reduction in the chemical cost of treatment up to and including the bleaching stage is equal to Rs. 35 per ton of pulp, and the less drastic conditions of digestion temperature which the process permits, results in a further gain of 2 per cent. in pulp yield.

4. The factory plant consists of crushing and cutting machines for the preparation of raw material for digestion, a pair of high pressure digesters coupled to four washing vessels or leachers, a beater carrying 4 cwt. of pulp and a drying machine, which may be used for either pulp or paper, capable of making a sheet 32 inches wide. The whole plant could produce, if continuously worked, 8 tons per week, involving the consumption of 20 tons of raw material. (See photograph opposite.)

5. This plant has cost with buildings about two and a half lakhs of rupees and the question may be pertinently asked whether Government is justified in sanctioning such an expenditure. The broad answer to this

is that Government owns unsaleable goods, for example, enormous bamboo forests, on which it desires to realise revenue and, to do so, must itself investigate their value, render them saleable, if possible, and advertise them. The objects of the Section are, first, to open up new supplies of raw materials to Indian paper-makers whose chief difficulty at present is the limited quantity and high cost of their present material as is evidenced by their considerable import of European woodpulp—and, second, to encourage the establishment of an export industry in pulp. A corollary to both is of course, the increase of forest revenues. If it is asked why the first object should not be prosecuted by the Indian paper-makers themselves the reply is that they are too few in number—there are only three or four firms of importance—and are not well organised scientifically for any such united and common effort, and they could not be expected to be interested in the second object while the effort to promote both objects costs Government no more than the prosecution of the second. Even in Canada and America where paper and pulp mills are numbered by the hundred, it has been found necessary by the Governments of both countries to establish their own Pulp Research Laboratories in the interests of the forests they control. To quote a paper-maker who has recently visited the experimental factory "The improved standing given to bamboo and grasses by the removal of the bleaching difficulty has more than justified the whole expenditure."

6. A considerable part of the activities of this Section lies in the exploration, in conjunction with district forest officers, of forest areas regarded as suitable for exploitation by this industry. The manufacturing facilities and transport problems associated with the raw material are fully explored and reported on to the local Governments concerned. Such a report has recently been published by the Industries Department of the Government of Bihar and Orissa as "Bulletin No. 5, Cuttack Bamboo Pulping Project." A similar scheme in the Tinnevelly Hills is now under consideration in conjunction with the Chief Forest Engineer to the Government of Madras and an exploration of the Kasalong areas near Chittagong led to the grant of a concession by the Government of Bengal which has resulted in the establishment of a bamboo papermill near Calcutta.

VII.

Minor Forest Products Section

BY

H. TROTTER, I.F.S.

The history of the Section of Minor Forest Products dates from the commencement of the Economic Branch. When Mr. R. S. Pearson, the present Forest Economist, first came to Dehra Dun over 15 years ago, he found it difficult to fill his time in with research work on timbers alone. There were in those days no Timber Testing Laboratory and no Wood Workshops or other facilities of this kind in the Institute and in consequence he turned his attention to Minor Forest Products. His pioneer work in this direction resulted in the establishment of several commercial undertakings dealing in minor products and a most useful foundation was laid for the further advancement of this important side of Indian Forestry. From the very start it was realised that paper pulp from bamboos and tanning materials from barks and leaves were two of the most likely avenues for profitable research and it was not long before separate Sections were inaugurated with an expert in charge, to make a detailed study of India's resources in these two products. The Section of Paper Pulp is dealt with in the previous chapter; the Section of Tans has been closed down for the present owing to the necessity for economy.

Two other lines of profitable research were, however, started in these early days, namely the rosin and turpentine industry in the United Provinces resulting in a steam distillation plant at Bhowali, near Naini Tal, and an investigation into the possibilities of the rosha grass oil industry which has resulted in eleven steam kilns now to be seen running in the rosha grass areas of the Central Provinces with a likelihood of a further extension of activities into other areas where the grass is found in any quantity*. At the same time a most valuable collection of Indian minor forest products was made and this collection is now on view in the museum of the Economic Branch at the Forest Research Institute at Dehra Dun and is well worth a visit by any one who has not seen it. (See photograph opposite page 33.)

These early days were, however, some of the brightest ever seen by this Section and since then it has passed through many vicissitudes. At the present day the Section of Minor Forest Products is unfortunately the Cinderella of the Economic Branch.

This is more to be regretted as India undoubtedly stands second to none in its forest wealth and organised control to keep in touch with

* A second rosha grass area has been leased this year to a firm starting work with 26 stills.

the trade and a very little practical research would give results which might be far reaching.

Unfortunately Dame Fortune stepped in at the wrong moment and the financial position of the Economic Branch during the last year or two has not been sufficiently strong to support an officer and the staff necessary to undertake this work. It may well be asked at the present time why minor products are being left to fend for themselves while so much is being done in research work on Timbers. The answer is merely a question of funds. The Economic Branch was allotted a fixed amount and was told to do the best it could with the money. While fully realising the importance of minor products funds would not admit of these, as well as timber, being properly treated. So, in preference to taking a half bite at the cherry, it was decided to make a really good show with the timber side, which after all is the *raison d'etre* of the Forest Department and to leave the Minor Products until such time as the subject could be taken up with full justice and a fair chance of success.

A forward movement was, however, made in 1922, when Mr. W. A. Robertson, a talented Forest Officer from Burma, held charge of the Section of Minor Forest Products for a short time. He at once started in to organise the section on sound lines and by the time he left the Institute about a year later, he had laid the foundation stone of a really practical Section. His first step was to get acquainted with the commercial possibilities and the supplies of minor forest products that were available. He also got into touch with likely markets and supply-contractors with a view to linking up the present gap between the Divisional Forest Officer and the commercial world. Unfortunately, owing to the reasons stated above, Mr. Robertson left the Institute when matters were looking promising and the post of Officer-in-Charge of Minor Forest Products has been held in abeyance ever since.

The Forest Economist and the writer have endeavoured to carry on as best they could, but any systematic research work is, of course, out of the question. To those unacquainted with India's wealth in minor forest products a few words of explanation may not be out of place.

In the first place there are about 5,000 different species of trees, climbers, shrubs and grasses scattered over the forests of India and the majority of these produce minor forest products. Naturally a great many of these products are not commercially valuable, but, on the other hand, there are a very large number which are of very real value, not only to India but to the outside world as well, and a great many more about which very little is known.

Take shellac for instance, the well known commercial product obtained from the lac insect, a minute creature about the size of a flea,

Photo by I. N. Sharma, May 1924.

Economic Museum.
Forest Research Institute, Dehra Dun, U.P.

[To face page 33].

which lives on the juices of certain forest trees and exudes a resinous substance from the pores of its body, with which it builds a protective covering over itself. Multiply the product of this one insect by several myriads and you have the world's supply of shellac. India holds a monopoly in the production of shellac and when one realises that the export value of this product from India last year amounted to no less than £7,250,000, one begins to think there may be something in this talk about minor forest products after all.

And this is one instance only, but it is also one of the instances where a considerable amount of research work has been done to help the industry and if the same policy was followed with regard to other products there is little doubt that the revenue of the country would very soon feel the benefit.

Amongst other products of importance which have established themselves in the commercial world are the well known **Distillation Products** of sandal wood oil, cutch and *katha*, rosha oil and lemon grass oil, *agar-attar*, camphor, charcoal, rosin and turpentine and many others.

Amongst the **Tans** the myrabolam fruit holds first place and over a million pounds sterling worth are exported from India annually. Several barks are of great value in tannage, such as those obtained from *babul* and *tarwad*. Of tan fruits, other than myrobalams, *babul* pods, *divi-divi* pods and a score of others could be mentioned.

As mentioned earlier in this note there used to be a separate Section of Tans under Mr. J. A. Pilgrim, who during the seven years that he held charge of the Section carried out some most useful investigations with regard to the tan possibilities of several areas in India. He started with the forests of Northern Bengal and continued with the tidal mangrove forests of the Sunderbans. He then turned his attention to Madras, finishing up with the oaks and chestnuts of North Burma and the tidal forests of South Tenasserim. Owing to the general slump in the tanning trade and shortage of funds at the Institute, the Tan Section had to be abolished in 1923.

To continue with the recital of our forest products we find amongst the **Dyes** the brilliant red kamela powder, santaline and *surgi* which are all seen to advantage in the bright coloured silks of the East.

Of greater importance are the **Gums and Resins** of which gum arabic, gum kino, the dammars and gurjan oil are only a few of the best known of a large variety.

Thitsi oil alone is responsible for having originated an art industry peculiar to Burma where it forms the basis of the well known Burmese lacquer ware.

Among the **Drugs** and **Spices** there are several which are of real vital importance not only to the medical world but to the human race itself. Podophyllin and aconite, strychnine and cardamoms are all well known but how many people realise that they are obtained to a large extent from the forests of India ?

In this connection it is interesting to note that medical research has lately discovered a cure for blackwater fever and " kala-azar, " in the leaves of *Vitex peduncularis*, a fairly common tree both in India and Burma. Hitherto no reliable cure was known for these two virulent fevers and this is a typical illustration of a valuable drug having remained untried for so many years for want of proper research.

Edible products also are a source of great relief to many a forest dweller and even a Forest Officer himself does not despise a succulent mangosteen, a Himalayan walnut, or the luscious wild asparagus found in the forests of Burma.

The **Bamboos, Canes and Grasses** are a study in themselves and many a cavalry-man has blessed his lanceshaft which was cut from the Indian male bamboo, while the rattan walking stick is used all over the world. Paper pulp from bamboo will eventually probably turn out the most profitable of all minor forest products but there is no need for an expansion of this theme here as the whole question has been dealt with in a separate note by the Officer-in-Charge, Paper Pulp Section. One could carry on with this list *ad infinitum*, even down to bees wax, honey and the edible bird's nests so palatable to the Chinese mandarin, but the object of this note is not to tire the reader but to interest him. He will have realised by now what the wealth of India's minor forest products means. It is no idle boast nor is it an impossible dream. India possesses the raw material and the world's markets are open to her.

Surely then it is the work of Forest Department to realise this wealth. Nature is very liberal and when dealing with minor forest products it is seldom necessary to cut down the tree so that any work connected with these products can be carried on in conjunction with the production of the major product (timber), and with little deterioration to the forest. The real fault, however, lies in the inability of the Divisional Officer to get into touch with the trade and *vice versâ*. The majority of the minor forest product industries are purely local at present. Leases or privilege rights are given year after year to small contractors and villagers, who carry on in exactly the same way as their great-grandfathers did and in many cases their methods are of the crudest description. The result of this is that genuine buyers on a large scale are seldom able to get into touch with these small suppliers and even if they do get in touch the result is very often unsatisfactory.

It is for this reason, therefore, that a central liaison officer, in the shape of an Officer-in-Charge of Minor Forest Products, would be invaluable. If he is able to organise matters so as to put the demand in touch with the supply his time will not have been wasted. And if he is able further, to investigate promising forest industries and to work

Old method of distilling Rosha Grass Oil.

New method of distillation of Rosha Grass Oil with copper steam stills.

[To face page 34].

up a trade in hitherto little exploited products he will have performed a very valuable duty to India.

No one has ever questioned the wealth of India's minor forest products but many have questioned India's lack of enterprise and it is up to the Forest Department to remove this blot from its escutcheon.

Meliora speramus is the motto of the Department and there was never a truer word spoken than when it was applied to the minor forest products of India.

VIII.

The Sub-Sections of Wood-Working, Veneers and Sawmill.

BY

H. TROTTER, I.F.S.

It was the original intention when the new Economic Branch was first outlined, to have three separate Sub-Sections for wood-working, veneers, and sawmill, with an upper grade assistant in charge of each.

This plan was later modified owing to financial and other reasons, and a European Superintendent was placed in charge of all three Sub-Sections and a European carpenter was appointed to superintend and help with the high class work required in the Wood Workshop itself. The latter has now returned to England and it is hoped that a suitable Indian trained in one of the wood-working institutes of this country will be appointed to replace him.

Wood-Workshops.

The duties of the wood-workshops are manifold and extend from preparing 18 feet beams for the Timber Testing Section to making insect boxes for the Entomological Branch.

First and foremost, however, is the work of preparing material for the other Sections and in this direction the wood-workshop is the source of the stream which continually feeds the Timber Testing Section and others with a perpetual supply of prepared material on which to experiment.

The Timber Testing Section for instance, now requires something in the neighbourhood of 18,000 specimens a year and the preparation of these alone is, therefore, no small achievement.

In addition to this work, all furniture and fittings for the new offices and workshops have been and will continue to be made in our own workshop and the furniture required for the new Main Building will also be made by our own carpenters.

An important side-line connected with this work is a careful observation of all timbers used. Woods of different kinds and different species are received from all over India and Burma and this was perceived to be a golden opportunity to compare their working qualities and suitability for various purposes.

Mr. W. Nagle, the Wood Workshop Superintendent, is himself well acquainted with a good many timbers of the world and his experience

Photo by I. N. Sharma.

Wood Workshop.

New Forest Research Institute, Dehra Dun, 1924.

[To face page 36].

Photo by I. N. Sharma.

Veneer Shop, showing veneer lathe, clipper and moving table.
New Forest Research Institute, Dehra Dun, 1924.

[To face page 37].

and practical knowledge is therefore invaluable in noting the working qualities of our Indian timbers on high speed planing, moulding and turning machines, and in watching how they behave when made up into high class furniture, etc.

Glue-testing and different methods of "building-up" joints to stand the adverse climatic conditions of this country are also carried out on scientific lines. In fact, as stated above, the duties of the wood workshop are manifold and to any one visiting the Institute they are also manifest.

The Wood Workshop Superintendent is always available to give advice on any subject connected with the working qualities of timber, the use of wood-work machinery or the care and treatment of tools and especially saws and any one who is in difficulties on these matters cannot do better than apply to the Forest Economist for expert advice. The Wood-Workshop Superintendent also gives a series of lectures on woodworking and the maintenance of tools and saws to Forest Students.

Veneers.

The sub-section of veneers has only just lately been brought into full running order but the veneers already turned out by this sub-section lead one to speak most optimistically on this subject. The plant consists of a rotary cutter which will peel logs up to 3 feet long, a clipper for cutting the peeled veneer into standard lengths, a glue-spreader, and a press for compressing the finished ply-wood. To anyone who has not seen a rotary cutter at work this method of peeling a log round its longitudinal axis is a most fascinating experience. The log is placed on the cutter as a smaller piece of wood is placed on a lathe and the knife peels off the wood round the log as it rotates, so that a continuous thin veneer comes away like an unrolled carpet. This is clipped into suitable lengths by the clipper and these veneer "boards", which can be cut from $\frac{1}{16}$th of an inch to $\frac{1}{4}$th of an inch in thickness according to requirements, are then dried in the seasoning kilns. After this they are glued by being passed through the gluing machines and are then stuck together in layers of 3, 7 or 9 thicknesses according to the type of ply-wood required.

The beautiful "figuring" obtained by a rotary cut on timbers from the temperate zones is well known and requires no description, but little is known of veneering many Indian hardwoods, so that much work still remains to be done in this country. As the result of the high hopes held out by the Economic Branch of the Forest Research Institute in the use of ply-wood for panelling, the Imperial Works Department of New Delhi have decided to panel the New Legislative Assembly and Council Chambers with three, five and nine ply-wood, as being the only means of constructing large handsome panels which would stand up against the adverse climatic conditions of Delhi.

As the result of some trial panels, made and sent to Delhi for inspection, every hope is held out that the work will be carried out satisfactorily, and the new Parliament Halls of Delhi will always remain as a reminder, not only to the Members themselves, but also to any one visiting the Chambers, that India not only possesses some extremely ornamental timbers, but that in the use of ply-wood lies the solution of many of India's troubles with regard to a really serviceable type of wood which will replace the solid woodwork now used which has failed so dismally in the past to stand up to the trying climatic conditions of this country.*

Sawmill.

A sawmill is naturally a very necessary adjunct to the Economic Branch of the Forest Research Institute. The sawmill at Kowlagarh consists of a log-pond about 120 feet long and 6 feet deep, an overhead moveable gantry which runs from the log-pond to the sawmill itself, a 52 inch inserted-tooth rip-saw with an automatic feed-rack, a 36 inch radial arm automatic feed re-saw, a pendulum cross-cut saw and a small trimming bench.

The whole mill is run with one 75 B. H. P. motor which will carry the full load of all machines running together.

On arrival at Dehra Dun all logs are at once barked and put into the log-pond. This prevents their cracking in the hot sun and keeps them in good condition until such time as the Section for which the timber is intended is ready to deal with it.

All logs which float are at once sunk and kept under water by means of iron rails fixed across the pond. When a log is required for conversion it is gripped by two clamp hooks attached to the over-head gantry, raised out of the pond by a pulley chain and the gantry carriage is then moved along the rails until it is over the log-roll, where the log is deposited. From there it is man-handled with peavies on to the saw-carriage and is ready for conversion into sizes required. The 52 inch inserted-tooth saw is naturally very wasteful in kerf and is only used to break down the log into convenient sizes which can pass through the radial arm saw for conversion into planks, scantlings, etc.

The pendulum cross-cut saw has been found most satisfactory for quick work, and has proved invaluable for cutting up paving-blocks from *chir* sleepers for use in all the workshops and godowns.

The sawmill is only a small one as sawmills go, but as it is not intended to be commercial in any way but purely to serve as a supply agent to the Sections it has been found satisfactory. The lay-out is good and it amply serves the purpose for which it was intended.

---

*At the time of going to Press two rooms have been completely panelled in plywood made from Andaman Padauk and Hazara walnut and other rooms will be completed shortly.

Photo by I. N. Sharma.

52 inch breaking down saw in sawmill.
New Forest Research Institute, Dehra Dun, 1924.

[To face page 38].

IX.

The Mechanical Sub-Section.

BY

H. TROTTER, I.F.S.

A word or two is perhaps necessary to explain the functions of the Mechanical Sub-Section which includes the Iron and Electrical workshops. In a large concern like the new Forest Research Institute and more especially the Economic Branch, there is always bound to be good deal of repair work and the maintenance of both the mechanical and electrical plant in good running order is the keystone of good work and good results. A stoppage of the air-circulation in the seasoning kilns, for instance, is enough to spoil the timber or at any rate the experimental data of that particular " run."

It is easily understood therefore, that with something like 20 electric motors, 30 complicated wood-working and timber-testing machines, a 150 B. H. P. Babcock and Wilcox boiler, in addition to the paper pulp, creosoting and seasoning plants, a small mechanical sub-section is a matter of vital importance, the more so as no mechanical engineering firm is to be found anywhere near Dehra Dun. This sub-section is in the charge of a Bombay trained Mechanical Engineer, Mr. Ram Das Tandan, with an Electrical Assistant and the necessary subordinate staff of fitters, blacksmiths, and boilermen to help him.

MGIPC—M—IV-5a-103—14-9-25—1,100.